Knots can only be tied in three-dimensional space. With fewer than three dimensions you cannot tie a knot, and with more than three dimensions any knot will always comes undone.

Cambridge U.K.
16-6-2017

First published 2000 AD
Revised edition © Wooden Books Ltd 2012 AD

Published by Wooden Books Ltd.
Glastonbury, Somerset

British Library Cataloguing in Publication Data
Watkins, M.
Useful Mathematical & Physical Formulæ

A CIP catalogue record for this mathemagical
gem is available from the British Library

ISBN 1 904263 00 3

Printed on 100% FSC approved sustainable papers
by RR Donnelley Asia Printing Solutions Ltd.

USEFUL
MATHEMATICAL & PHYSICAL
FORMULÆ

compiled by
Matthew Watkins

and illuminated by
Matt Tweed

Thanks and praises to Jah! Rastafari!

This book is dedicated to Mum and Dad, and to Inge,
who showed me that the really important things
can't possibly be described with formulas.

For further reading I recommend
"Time and Number" by Marie-Louise von Franz,
and "A Beginner's Guide to Constructing the Universe:
The Mathematical Archetypes of Nature, Art, and Science"
by Michael S. Schneider.

Contents

The Equations of Quantum Mechanics

Planck - Energy of a Quantum

$$E = h \cdot f$$

h = Planck's constant

Speed of Light

$$c = f\lambda$$

f = frequency

de Broglie - Wavelength

$$\lambda = h / p$$

Momentum

$$p = mc$$

m=mass

$$i = \sqrt{-1}$$

Wave number

$$\frac{2\pi}{\lambda} = \frac{p}{\hbar}$$

Einstein - Energy

$$E = mc^2$$
$$= pc$$

Kinetic Energy

$$E = \tfrac{1}{2}mc^2 = \frac{p^2}{2m}$$

Dirac's Constant

$$\hbar = \frac{h}{2\pi}$$

Angular Velocity

$$\omega = 2\pi f = \frac{Et}{\hbar}$$

Wave function in Classical Mechanics

$$\psi = A\cos\left(\frac{2\pi r}{\lambda} - \omega t\right)$$

Euler's Equation

$$e^{i\theta} = \cos\theta + i\sin\theta$$

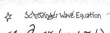

Schrödinger Wave Equation

$$i\hbar\frac{\partial}{\partial t}\psi(x,y,z,t) = \left(-\frac{\hbar^2}{2m}\nabla^2 + V(x,y,z)\right)\psi(x,y,z,t)$$

Energy Operator   Kinetic energy + potential energy   gives probability

Hamiltonian Operator

Simplified Form

$$\hat{E}\psi = \hat{H}\psi$$

∇^2 is the Laplacian operator $\dfrac{\partial^2}{\partial x^2} + \dfrac{\partial^2}{\partial y^2} + \dfrac{\partial^2}{\partial z^2}$

INTRODUCTION

This little book is an attempt to present the basic formulæ of mathematics and physics in a friendly and usable form.

The use of number and symbol to model, predict, and manipulate reality is a powerful form of sorcery. Unfortunately, possession of these abilities doesn't necessarily bring with it wisdom or foresight. As a result, we see the proliferation of dangerous technologies and an increasing obsession with *quantity*, typified by the subservience of almost everything to the global economy. Readers are advised to use the contents of this book with due care and attention.

On the other hand, mathematical tools have made it possible to perceive unity in seemingly separate areas. Light and electricity, for example, once unrelated topics, are now both understood in terms of the theory of *electromagnetic fields*.

Brilliantly illustrating this 'double-edged sword', Einstein's well-known $E = mc^2$, perhaps the most famous formula of all, will be forever linked to both the creation of atomic weapons and the scientific (re)discovery of the unity of matter and energy.

May your senses of awe and delight never be extinguished!

TRIANGLES
and their various centres

A *right-angled* triangle obeys *Pythagoras' Law*: the square of the *hypotenuse* (the side opposite the right angle) is equal to the sum of the squares of the other two sides (*opposite top left*).

$$a^2 + b^2 = c^2 \quad \text{or equivalently} \quad c = \sqrt{a^2 + b^2}$$

The sum of internal angles in any triangle = 180°, or π *radians*.

The *perimeter* $P = a + b + c$

The area $S = \frac{1}{2}bh = \frac{1}{2}ab\sin C$ (*see opposite top right*).

Sine Rule: $\dfrac{a}{\sin A} = \dfrac{b}{\sin B} = \dfrac{c}{\sin C} = 2r$

(where r is the radius of the *circumcircle*).

Medians connect corners to the midpoints of the opposite sides. The three medians meet at a point called the *centroid*:

$$m^a = \frac{1}{2}\sqrt{2(b^2 + c^2) - a^2} \qquad m^b = \frac{1}{2}\sqrt{2(a^2 + c^2) - b^2}$$

$$m^c = \frac{1}{2}\sqrt{2(a^2 + b^2) - c^2}$$

Altitudes are line segments drawn from each corner to the opposite side (or its extension), meeting it at a right angle:

$$h_a = \frac{2S}{a} \qquad h_b = \frac{2S}{b} \qquad h_c = \frac{2S}{c}$$

The three altitudes meet at the *orthocentre*.

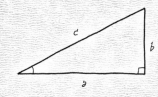

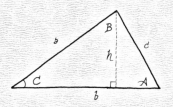

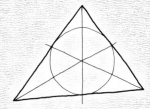

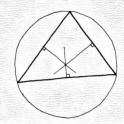

a triangle's angular bisectors meet at the centre of its incircle

a triangle's perpendicular bisectors meet at the centre of its circumcircle

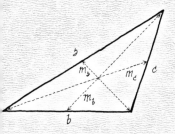

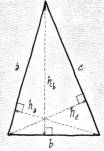

...ians are drawn to the midpoints of each side and give the centre of gravity of a uniform sheet

altitudes intersect at the orthocentre (not necessarily inside the triangle)

Two-Dimensional Figures
areas and perimeters

Formulæ for the perimeters and areas of various two-dimensional forms are given below.

Circle: Radius = r, diameter $d = 2r$
Perimeter, or *circumference* = $2\pi r = \pi d$
Area = πr^2 where $\pi = 3.1415926\ldots$

Ellipse: Area = $\pi a b$
a and b are the *minor* and *major semi-axes* respectively. The two illustrated points are its *foci*, such that $l + m$ is constant.

Rectangle: Area = ab
Perimeter = $2a + 2b$

Parallelogram: Area = $bh = ab \sin \alpha$
Perimeter = $2a + 2b$

Trapezium: Area = $\frac{1}{2} h (a + b)$
Perimeter = $a + b + h \left(\csc \alpha + \csc \beta \right)$

Regular n-gon: Area = $\frac{1}{4} n b^2 \cot (180°/n)$
Perimeter = nb
Sides and internal angles are all equal.

Quadrilateral (i): Area = $\frac{1}{2} ab \sin \alpha$

Quadrilateral (ii): Area = $\frac{1}{2} (h_1 + h_2) b + \frac{1}{2} a h_1 + \frac{1}{2} c h_2$

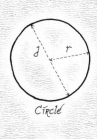

Circle

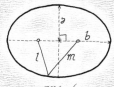

Ellipse

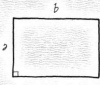

Rectangle

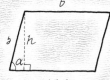

Parallelogram

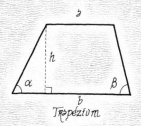

Trapezium

Regular n-sided polygon

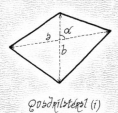

Quadrilateral (i)

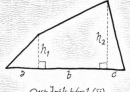

Quadrilateral (ii)

THREE-DIMENSIONAL FIGURES
volumes and surface areas

Formulæ for the volumes and surface areas (including bases) of eight three-dimensional solids are given below.

Sphere:
Volume = $\frac{4}{3}\pi r^3$
Surface area = $4\pi r^2$

Box:
Volume = abc
Surface area = $2(ab + ac + bc)$

Cylinder:
Volume = $\pi r^2 h$
Surface area = $2\pi rh + 2\pi r^2$

Cone:
Volume = $\frac{1}{3}\pi r^2 h$
Surface area = $\pi r\sqrt{r^2 + h^2} + \pi r^2$

Pyramid:
Base area A
Volume = $\frac{1}{3}Ah$

Frustum:
Volume = $\frac{1}{3}\pi h(a^2 + ab + b^2)$
Surface area = $\pi(a + b)c + \pi a^2 + \pi b^2$

Ellipsoid:
Volume = $\frac{4}{3}\pi abc$

Torus:
Volume = $\frac{1}{4}\pi^2(a + b)(b - a)^2$
Surface area = $\pi^2(b^2 - a^2)$

Sphere

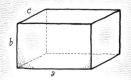

Box (Rectangular parallelepiped)

Cylinder

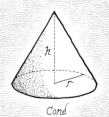

Cone

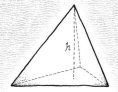

Pyramid with polygonal base, area A

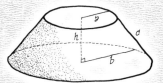

Frustum (truncated cone)

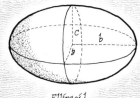

Ellipsoid

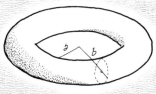

Torus

COORDINATE GEOMETRY
axes, lines, gradients and intersections

A pair of axes imposed on the plane at right angles allow a point to be defined by a pair of real numbers (*opposite*). The axes intersect at $(0,0)$, the *origin*. Horizontal and vertical positions are often referred to as x and y respectively.

The equation of a line is given by $y = mx + c$ where m is its gradient. This line cuts the y-axis at $(0, c)$ and the x-axis at $(-\frac{c}{m}, 0)$.

A vertical line has a constant x value, taking the form $x = k$.

The line passing through (x_0, y_0) with gradient n is given by the equation $y = nx + (y_0 - nx_0)$. A line perpendicular to another of gradient n will have gradient $-\frac{1}{n}$.

The equation of the line through (x_1, y_1) and (x_2, y_2) is,

$$y = \left(\frac{y_2 - y_1}{x_2 - x_1} \right) (x - x_2) + y_2 \quad \text{when } x_1 \neq x_2$$

The angle θ between two lines, gradients m and n, satisfies,

$$\tan \theta = \frac{m - n}{1 + mn}$$

A circle, radius r, centre (a, b) is given by $(x - a)^2 + (y - b)^2 = r^2$.

In three dimensions, a z-axis is added and many equations take analogous forms. For instance, a sphere with radius r and centred at (a, b, c) is given by $(x - a)^2 + (y - b)^2 + (z - c)^2 = r^2$. The general equation for a three-dimensional plane is $ax + by + cz = d$.

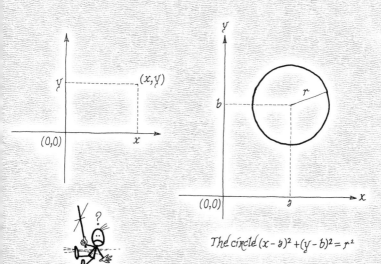

The circle $(x-a)^2 + (y-b)^2 = r^2$

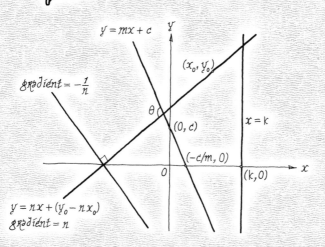

$y = mx + c$

gradient $= -\dfrac{1}{n}$

(x_0, y_0)

θ

$(0, c)$

$x = k$

$(-c/m, 0)$

$(k, 0)$

$y = nx + (y_0 - nx_0)$
gradient $= n$

TRIGONOMETRY
applied to right triangles

A right triangle with sides of length *a, b, c,* and angle θ is shown opposite left. The six *trigonometric functions*: sine, cosine, tangent, cosecant, secant, and cotangent are then defined as follows:

$$\sin\theta = \frac{b}{c} \qquad \cos\theta = \frac{a}{c} \qquad \tan\theta = \frac{b}{a}$$

$$\csc\theta = \frac{c}{b} \qquad \sec\theta = \frac{c}{a} \qquad \cot\theta = \frac{a}{b}$$

The sine and cosine are the height and base of a right triangle in a circle of radius 1 as shown below and opposite top right:

$$a = \cos\theta \quad \text{and} \quad b = \sin\theta$$

By Pythagoras' Theorem (*see page 2*) we know $a^2 + b^2 = c^2$, hence for any angle θ in the circle we have the important identity,

$$\cos^2\theta + \sin^2\theta = 1$$

Sines, cosines and tangents have positive or negative values in different quadrants of the circle as shown below right.

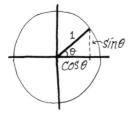
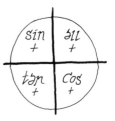

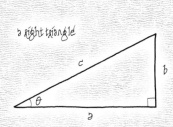

a right triangle

$a = c \cos \theta$
$\quad = b \cot \theta$

$b = c \sin \theta$
$\quad = a \tan \theta$

$c = a \sec \theta$
$\quad = b \csc \theta$

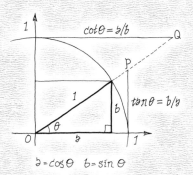

$\cot \theta = a/b$

$\tan \theta = b/a$

$a = \cos \theta \quad b = \sin \theta$

tangents and cotangents as lengths.
OP has length $\sec \theta$ and OQ has length $\csc \theta$.

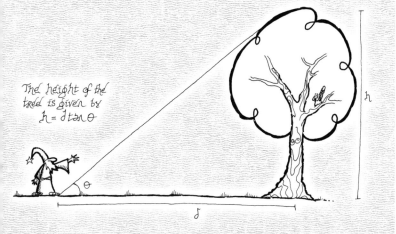

The height of the
tree is given by
$h = d \tan \theta$

TRIGONOMETRIC IDENTITIES
relating the six functions

The definitions on the previous page lead to the following:

$$\tan\theta = \frac{\sin\theta}{\cos\theta}, \quad \cot\theta = \frac{\cos\theta}{\sin\theta}, \quad \sec\theta = \frac{1}{\cos\theta}, \quad \csc\theta = \frac{1}{\sin\theta}$$

Dividing $\cos^2\theta + \sin^2\theta = 1$ by $\cos^2\theta$ and $\sin^2\theta$ produces:

$$1 + \tan^2\theta = \sec^2\theta \qquad \text{and} \qquad 1 + \cot^2\theta = \csc^2\theta$$

The six functions applied to negative angles produce:

$$\sin(-\theta) = -\sin\theta \qquad \cos(-\theta) = \cos\theta \qquad \tan(-\theta) = -\tan\theta$$

$$\csc(-\theta) = -\csc\theta \qquad \sec(-\theta) = \sec\theta \qquad \cot(-\theta) = -\cot\theta$$

Angle sum formulæ apply when two angles are combined:

$$\sin(\alpha + \beta) = \sin\alpha\cos\beta + \cos\alpha\sin\beta$$

$$\cos(\alpha + \beta) = \cos\alpha\cos\beta - \sin\alpha\sin\beta$$

$$\tan(\alpha + \beta) = \frac{\tan\alpha + \tan\beta}{1 - \tan\alpha\tan\beta}$$

For doubled or tripled angles, use *multiple angle formulæ*:

$$\sin 2\alpha = 2\sin\alpha\cos\alpha \qquad\qquad \sin 3\alpha = 3\sin\alpha\cos^2\alpha - \sin^3\alpha$$

$$\cos 2\alpha = \cos^2\alpha - \sin^2\alpha \qquad\qquad \cos 3\alpha = \cos^3\alpha - 3\sin^2\alpha\cos\alpha$$

$$\tan 2\alpha = \frac{2\tan\alpha}{1 - \tan^2\alpha} \qquad\qquad \tan 3\alpha = \frac{3\tan\alpha - \tan^3\alpha}{1 - 3\tan^2\alpha}$$

The graph opposite is plotted in radians (π radians = 180°). A rough lookup table is also shown.

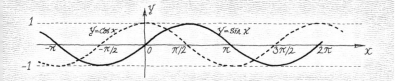

Radians	Degrees	Sine	Cosine	Tangent	Secant	Cosecant	Cotangent
0	0°	0	1	1	∞	1	∞
	2.5°	0.04362	0.9990	0.04366	22.926	1.00095	22.904
	5°	0.08716	0.9962	0.08749	11.474	1.00382	11.430
	7.5°	0.1305	0.9914	0.1317	7.6613	1.00863	7.5958
	10°	0.1736	0.9848	0.1763	5.7588	1.0154	5.6713
	12.5°	0.2164	0.9763	0.2217	4.6202	1.0243	4.5107
	15°	0.2588	0.9659	0.2679	3.8637	1.0353	3.7321
	17.5°	0.3007	0.9537	0.3153	3.3255	1.0485	3.1716
	20°	0.3420	0.9997	0.3640	2.9238	1.0642	2.7475
π/8	22.5°	0.3827	0.9239	0.4142	2.6131	1.0824	2.4142
	25°	0.4226	0.9063	0.4663	2.3662	1.1034	2.1445
	27.5°	0.4617	0.8870	0.5206	2.1657	1.1274	1.9210
	30°	0.5	0.8660	0.5774	2	1.1547	1.7321
	32.5°	0.5373	0.8434	0.6371	1.8612	1.1857	1.5697
	35°	0.5736	0.8192	0.7002	1.7434	1.2208	1.4281
	37.5°	0.6088	0.7934	0.7673	1.6427	1.2605	1.3032
	40°	0.6428	0.7660	0.8391	1.5557	1.3054	1.1918
	42.5°	0.6756	0.7373	0.9163	1.4802	1.3563	1.0913
π/4	45°	0.7071	0.7071	1	1.4142	1.4142	1
	47.5°	0.7373	0.6756	1.0913	1.3563	1.4802	0.9163
	50°	0.7660	0.6428	1.1918	1.3054	1.5557	0.8391
	52.5°	0.7934	0.6088	1.3032	1.2605	1.6427	0.7673
	55°	0.8192	0.5736	1.4281	1.2208	1.7434	0.7002
	57.5°	0.8434	0.5373	1.5697	1.1857	1.8612	0.6371
	60°	0.8660	0.5	1.7321	1.1547	2	0.5774
	62.5°	0.8870	0.4617	1.9210	1.1274	2.1657	0.5206
	65°	0.9063	0.4226	2.1445	1.1034	2.3662	0.4663
3π/8	67.5°	0.9239	0.3827	2.4142	1.0824	2.6131	0.4142
	70°	0.9397	0.3420	2.7475	1.0642	2.9238	0.3640
	72.5°	0.9537	0.3007	3.1716	1.0485	3.3255	0.3153
	75°	0.9659	0.2588	3.7321	1.0353	3.8637	0.2679
	77.5°	0.9763	0.2164	4.5107	1.0243	4.6202	0.2217
	80°	0.9848	0.1736	5.6713	1.0154	5.7588	0.1763
	82.5°	0.9914	0.1305	7.5958	1.00863	7.6613	0.1317
	85°	0.9962	0.08716	11.430	1.00382	11.474	0.08749
	87.5°	0.9990	0.04326	22.903	1.00095	22.926	0.04366
π/2	90°	1	0	∞	1	∞	1

Spherical Trigonometry
formulæ for heaven and earth

A *spherical triangle* has internal angles whose sum is between 180° and 540°. Its sides are arcs of *great circles* (whose centres all lie at the sphere's centre). Any two points on a sphere can define a *great circle*, and any three a *lesser circle*. Any circle on a sphere, greater or lesser, defines two *poles*.

The sides of a spherical triangle can be thought of as angles. The six relevant quantities are shown opposite and obey:

Law of Sines: $\dfrac{\sin a}{\sin A} = \dfrac{\sin b}{\sin B} = \dfrac{\sin c}{\sin C}$

Law of Cosines: $\cos a = \cos b \cos c + \sin b \sin c \cos A$

$\cos A = -\cos B \cos C + \sin B \sin C \cos a$

Law of Tangents: $\dfrac{\tan\frac{1}{2}(A+B)}{\tan\frac{1}{2}(A-B)} = \dfrac{\tan\frac{1}{2}(a+b)}{\tan\frac{1}{2}(a-b)}$

Spherical trigonometry is used in navigation. For example, using degrees *longitude* and *latitude*, a ship sails from Q to R:

$$a = 90° - \text{lat. } R \qquad b = 90° - \text{lat. } Q \qquad C = \text{long. } Q - \text{long. } R$$

C is known as the *polar angle*. The initial and final courses are given by A and B and the length of the journey by c. Use the following equations to solve for B, A, and c:

$$\tan\tfrac{1}{2}(B+A) = \cos\tfrac{1}{2}(b-a) \; \sec\tfrac{1}{2}(b+a) \; \cot\tfrac{1}{2}C$$

$$\tan\tfrac{1}{2}(B-A) = \sin\tfrac{1}{2}(b-a) \; \csc\tfrac{1}{2}(b+a) \; \cot\tfrac{1}{2}C$$

$$\tan\tfrac{1}{2}c = \tan\tfrac{1}{2}(b-a) \; \sin\tfrac{1}{2}(B+A) \; \csc\tfrac{1}{2}(B-A)$$

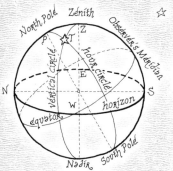

Celestial great circles

The Astronomical triangle PZT, for some celestial object

side TZ = zenith distance of T = 90° - altitude of T
side TP = polar distance of T = 90° - declination of T
side ZP = 90° - or + latitude of observer (N. or S. hemisphere)
angle PZT = azimuth of T (if T is east of observer's meridian)
    or 360° - azimuth of T (if T is west of OM)
angle ZPT = hour angle of T (if T is west of OM) in hours ' "
    or 360° - hour angle (if T is east of OM)

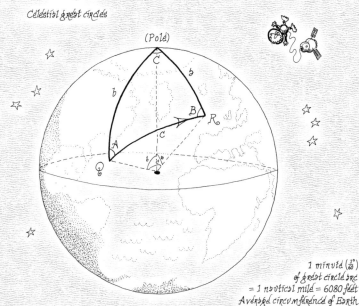

1 minute ($\frac{1}{60}$°)
of great circle arc
= 1 nautical mile = 6080 feet
Average circumference of Earth
= 24860 miles.

THE QUADRATIC FORMULA
discriminants and parabolas

A *quadratic equation* is one of the form $ax^2 + bx + c = 0$ (where a is not zero). The solutions, or *roots*, of such an equation are given by the quadratic formula,

$$\frac{-b \pm \sqrt{b^2 - 4ac}}{2a}$$

The quantity $b^2 - 4ac$ is called the *discriminant* which dictates the nature and number of the solutions. There are three cases:

$b^2 - 4ac > 0$ two different (real) solutions

$b^2 - 4ac = 0$ only one (real) solution

$b^2 - 4ac < 0$ two different *complex* or *imaginary* solutions
                      (as opposed to *real* ones—*see page 112*)

Examples (*shown opposite*):

i) $2x^2 - x - 1 = 0$ has a discriminant of 9 and produces two real solutions, 1 and $-\frac{1}{2}$.

ii) $x^2 - 2x + 1 = 0$ yields a discriminant of zero and hence has one real root, namely $x = 1$.

iii) $4x^2 + 8x + 5 = 0$ is an example of a quadratic equation with no real roots, solving for $x = -1 + \frac{i}{2}$ and $x = -1 - \frac{i}{2}$ (*see page 112*).

The quadratic equation $ax^2 + bx + c = 0$ has real solutions at values of x where the graph of the function $y = ax^2 + bx + c$ crosses the x-axis (*i.e.* where $y = 0$).

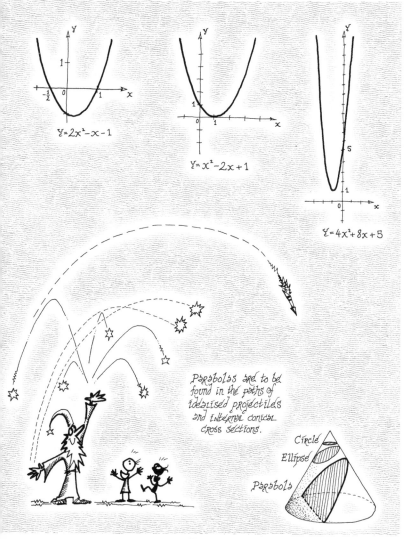

$y = 2x^2 - x - 1$

$y = x^2 - 2x + 1$

$y = 4x^2 + 8x + 5$

Parabolas are to be found in the paths of idealised projectiles and internal conical cross sections.

Circle
Ellipse
Parabola

MATRICES AND VECTORS
basic operations and determinants

An $m \times n$ matrix is a rectangular array of numbers with m rows and n columns. Two $m \times n$ matrices can be added or subtracted in the obvious way, and any matrix can be similarly multiplied by a single number.

You can only multiply an $m \times n$ matrix by a $p \times q$ matrix if $n = p$ (i.e. the number of columns of the first equals the rows of the second). The procedure for matrix multiplication is shown below. Note that matrix multiplication is not commutative: AB is not necessarily equal to BA. In fact, BA is not defined here unless $m = q$. A *square* matrix has an equal number of rows and columns. The *determinant* $|A|$ of a square matrix A is an important scalar quantity associated with A (*see examples opposite*). Determinants of larger square matrices are defined iteratively.

The *inverse* of an $n \times n$ matrix A with $|A| \neq 0$ is an $n \times n$ matrix, A^{-1}, such that $AA^{-1} = A^{-1}A = I_n$ (the $n \times n$ identity matrix defined opposite).

Vectors describe *displacements* in space, e.g. $\mathbf{a} = (1, 1, 2)$. They are summed in the obvious way, with a closed circuit summing to a zero vector. The *dot* or *scalar product* of two vectors $\mathbf{a} \cdot \mathbf{b}$ is $\|\mathbf{a}\| \|\mathbf{b}\| \cos \theta$, where $\|\mathbf{a}\|$ is the length of the vector $\mathbf{a} = (x, y, z)$, e.g. $\sqrt{x^2 + y^2 + z^2}$, and θ is the angle between the vectors. The *cross product* of two three-dimensional vectors $\mathbf{a} \times \mathbf{b}$ is a vector perpendicular to both, $\|\mathbf{a}\| \|\mathbf{b}\| \sin \theta \, \mathbf{n}$, where $\mathbf{n}$ is the right-handed unit vector perpendicular to both.

$$\begin{pmatrix} a & b & c \\ d & e & f \end{pmatrix} \begin{pmatrix} p & s \\ q & t \\ r & u \end{pmatrix} = \begin{pmatrix} (ap + bq + cr) & (as + bt + cu) \\ (dp + eq + fr) & (ds + et + fu) \end{pmatrix}$$

identity matrices

$$\begin{bmatrix} 1 & 0 \\ 0 & 1 \end{bmatrix} \begin{bmatrix} 1 & 0 & 0 \\ 0 & 1 & 0 \\ 0 & 0 & 1 \end{bmatrix} \begin{bmatrix} 1 & 0 & 0 & 0 \\ 0 & 1 & 0 & 0 \\ 0 & 0 & 1 & 0 \\ 0 & 0 & 0 & 1 \end{bmatrix}$$

etc.

the transpose of a matrix

$$\begin{bmatrix} a & b & c \\ d & e & f \end{bmatrix} \text{ is } \begin{bmatrix} a & d \\ b & e \\ c & f \end{bmatrix}$$

the determinant of matrix

$$\begin{vmatrix} a & c \\ b & d \end{vmatrix} = \text{area of}$$

is $ad - bc$

the determinant of matrix

$$\begin{vmatrix} a & d & g \\ b & e & h \\ c & f & i \end{vmatrix} \text{ is } = \text{ volume of}$$

$aei + dhc + bfg$
$- ceg - bdi - fha$

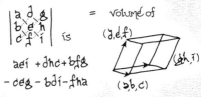

$$\begin{bmatrix} -1 & 0 \\ 0 & 1 \end{bmatrix}$$

reflection

$$\begin{bmatrix} 1 & 1.25 \\ 0 & 1 \end{bmatrix}$$

shearing

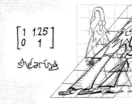

$$\begin{bmatrix} \tfrac{3}{2} & 0 \\ 0 & \tfrac{2}{3} \end{bmatrix}$$

scaling

$$\begin{bmatrix} \cos 30° & -\sin 30° \\ \sin 30° & \cos 30° \end{bmatrix}$$

rotation

EXPONENTIALS AND LOGARITHMS
growth and decay

Given some value a, we can define *a squared* and *a cubed* as follows: $a^2 = a \times a$, $a^3 = a \times a \times a$. In the expression a^n, n is the *exponent*. Here are the basic exponential formulæ:

$$a^0 = 1 \qquad\qquad a^p a^q = a^{p+q} \qquad\qquad (ab)^p = a^p b^p$$

$$a^{1/n} = \sqrt[n]{a} \qquad\qquad (a^p)^q = a^{pq} \qquad\qquad a^{m/n} = \sqrt[n]{a^m}$$

$$a^{-p} = \frac{1}{a^p} \qquad\qquad \sqrt[n]{\frac{a}{b}} = \frac{\sqrt[n]{a}}{\sqrt[n]{b}} \qquad\qquad \frac{a^p}{a^q} = a^{p-q}$$

The *base a logarithm* of x, $\log_a x = y$ is the quantity which satisfies $a^y = x$. Because $a^0 = 1$ and $a^1 = a$, then we always have both $\log_a a = 1$ and $\log_a 1 = 0$. Here are the essential logarithmic formulæ:

$$\log_a xy = \log_a x + \log_a y \qquad\qquad \log_a \frac{x}{y} = \log_a x - \log_a y$$

$$\log_a x^k = k \log_a x \qquad\qquad \log_a \frac{1}{x} = -\log_a x$$

$$\log_a \sqrt[n]{x} = \frac{1}{n} \log_a x \qquad\qquad \log_k a = \log_m a \, \log_k m$$

Any positive base (except 1) can be used, but most common are 10 and the constant e (= 2.718...), which occurs widely throughout the natural world, often in processes of growth and decay. $\log_e$ is usually just written *log* or *ln*. Logarithms allow multiplication and division of numbers by the addition or subtraction of exponents.

Exponentials are also useful for calculating compound growth. A quantity x which increases or decreases by p percent in unit time will, after time t take the values $x\left(1 + \frac{p}{100}\right)^t$ and $x\left(1 - \frac{p}{100}\right)^t$ respectively.

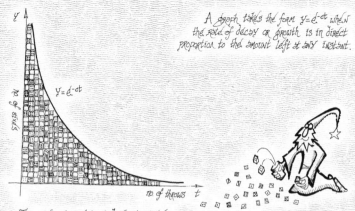

A graph takes the form $y = e^{-ct}$ when the rate of decay or growth is in direct proportion to the amount left at any instant.

$y = e^{-ct}$

no. of sixes

no. of throws t

Throwing y number of dice, stack the sixes. Rethrow the remainder, stack the sixes and repeat.

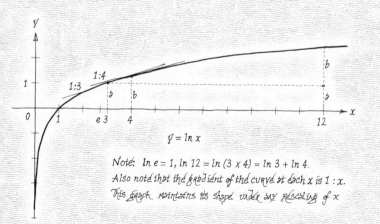

$y = \ln x$

Note: $\ln e = 1$, $\ln 12 = \ln (3 \times 4) = \ln 3 + \ln 4$.
Also note that the gradient of the curve at each x is $1 : x$.
This graph maintains its shape under any rescaling of x.

MEANS AND PROBABILITIES

safe proportions and risky outcomes

Given two numbers, *a* and *b*, then the three important averages or *means,* traditionally used in geometry, music, and architecture, are:

the *arithmetic mean,* $\frac{1}{2}(a + b)$,   the *geometric mean,* $\sqrt{ab}$

and the *harmonic mean,* $\frac{2ab}{a + b}$

Suppose we have a situation with *n* possible, equally likely outcomes, *k* of these being desired. The probability *p* of a desired outcome occurring on any one occasion is then,

$$p = \frac{number\ of\ desired\ outcomes}{total\ number\ of\ possible\ outcomes} = \frac{k}{n}$$

Note that *p* is always between 0 and 1. Imagine *E* and *F* are two independent possible events, with probabilities $P(E)$ and $P(F)$ of occurring, respectively. The probability of both *E and F* occurring is $(EF) = P(E) \times P(F)$. With this, we can express the probability of *either E or F* occurring as $P(E + F) = P(E) + P(F) - P(EF)$.

F is *not* independent of *E* when the probability of *F* occurring is altered by *E* occurring. In this case we define the *conditional probability* $P(F|E)$ to be the probability of *F* occurring once *E* has occurred, and we have $P(EF) = P(E) \times P(F|E)$. For example, if G and H both represent the choice of a black ball from the bag on our left, then $P(GH) = \frac{2}{5} \times \frac{1}{4} = \frac{1}{10}$ *(see page 56 for fractions).*

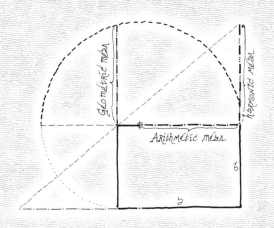

We are to choose one ball from each bag:

$P(E)$ = probability of choosing black ball from left bag = $\frac{2}{5}$

$P(F)$ = probability of choosing black ball from right bag = $\frac{4}{6} = \frac{2}{3}$

$P(EF)$ = probability of choosing black balls from both bags = $\frac{2}{5} \times \frac{2}{3} = \frac{4}{15}$

$P(E+F)$ = probability of choosing at least one black ball = $\frac{2}{5} + \frac{2}{3} - \frac{4}{15} = \frac{4}{5}$

COMBINATIONS & PERMUTATIONS
ways of arranging things

Suppose we have *n* items and we wish to consider groupings of *k* of them. Two types of groupings exist: *combinations*, where the order is not important, and *permutations*, where the order does matter. We need to use *factorials* here. The factorial of *m*, written *m*! (where $m \geq 1$) is defined as,

$$m! = m(m-1)(m-2) \ldots 2 \times 1 \ (e.g.\ 6! = 6 \times 5 \times 4 \times 3 \times 2 \times 1 = 720)$$

0 is a special case: $0! = 1$ by convention.

The number of combinations of *k* items out of *n* is then,

$$C_k^n = \frac{n!}{(n-k)\,k!}, \text{ also written } \binom{n}{k}$$

The (obviously larger) number of permutations is,

$$P_k^n = \frac{n!}{(n-k)} = k!\binom{n}{k}$$

In a situation having two possible outcomes, *P* and *Q*, with probabilities *p* and *q*, we know $p+q = 1$, so $(p+q)^n = 1$. The term $\binom{n}{k}p^{n-k}q^k$ in the *binomial expansion* of $(p+q)^n$ is then the probability of $n-k$ occurrences of *P* and *k* occurrences of *Q* from a total of *n*. The general binomial formula is,

$$(x+y)^n = x^n + \binom{n}{1}x^{n-1}y + \ldots + \binom{n}{k}x^{n-k}y^k + \ldots + \binom{n}{n-1}xy^{n-1} + \binom{n}{n}y^n$$

The rows of *Pascal's Triangle* correspond here. For example:

$$(x+y)^4 = x^4 + 4x^3y + 6x^2y^2 + 4xy^3 + y^4$$

Example:

$$C^3_2 = \frac{3!}{1!\,2!}$$

$$= \frac{6}{1\cdot2} = 3$$

Example:

$$P^3_2 = \frac{3!}{(3-2)!} = \frac{6}{1} = 6$$

```
            1
          1   1
        1   2   1
      1   3   3   1
    1   4   6   4   1
  1   5   10  10  5   1
1   6   15  20  15  6   1
```

Pascal's Triangle - in which each number is the sum of the two numbers above it. Row $n+1$ corresponds to the terms in the expansion of $(x + y)^n$.

STATISTICS
distribution and deviation

Statistical analysis allows us to process samples of observed data in order to reveal trends and make predictions. If $x_1, x_2, \ldots, x_n$ is a set of values of some measurable phenomenon, the average or *mean* value is given by $x = \frac{1}{n}(x_1 + \ldots + x_n)$.

The *standard deviation* σ of the sample then gauges the extent to which the data deviate from this mean:

$$\sigma = \sqrt{\frac{x_1^2 + \ldots + x_n^2}{n} - \bar{x}^2} = \sqrt{\frac{(x_1^2 - \bar{x}^2) + \ldots + (x_n^2 - \bar{x}^2)}{n}}$$

The most commonly occurring form of *statistical distribution* is the *normal* or *Gaussian*. Its general form produces a bell-shaped curve centred at x whose 'width' is dependent on σ:

$$f(x) = \frac{1}{\sigma\sqrt{2\pi}}\, e^{\frac{(x-x)^2}{2\sigma^2}}$$

In data which is *normally distributed* the probability of a value occurring in the range between a and b is equal to the area under the curve between these values as shown opposite. The total area under the curve (all possibilities) is equal to one.

The *Poisson distribution* tells us that if the mean number of events of a particular type in a fixed time interval is μ, then the probability of n events occurring in one interval is given by,

$$p(n) = \frac{\mu^n e^{-\mu}}{n!} \text{ where } e = 2.718\ldots \text{ (see page 52)}.$$

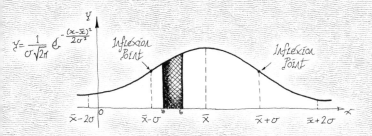

$$y = \frac{1}{\sigma\sqrt{2\pi}} \, \mathcal{e}^{-\frac{(x-\bar{x})^2}{2\sigma^2}}$$

Inflexion Point

Inflexion Point

$\bar{x}-2\sigma$ $\bar{x}-\sigma$ a b $\bar{x}$ $\bar{x}+\sigma$ $\bar{x}+2\sigma$

The standard deviation σ becomes a convenient unit by which to scale the sample range. Note the positions of the inflective points.

68.26 % of sample lies within σ of $\bar{x}$.

95.44 % of sample lies within 2σ of $\bar{x}$

99.73 % of sample lies within 3σ of $\bar{x}$.

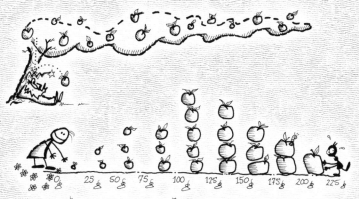

0 25 50 75 100 125 150 175 200 225

A random sample of apples arranged vertically according to a system of equal size brackets produces a frequency distribution graph. Given

a large enough sample, such a graph would begin to approximate a continuous curve ~ the graph of a normal (or Gaussian) distribution

KEPLER'S AND NEWTON'S LAWS
bodies in motion

Johannes Kepler (1571–1630) discovered three laws describing planetary motion. They hold true for all orbiting bodies in space.

1. *Planets move in ellipses with the sun at one focus.*

2. *A line drawn from the sun to the planet sweeps out equal areas in equal times.*

3. *The square of the time which a planet takes to orbit the Sun divided by the cube of its major semi-axis (see page 62) is a constant throughout the solar system.*

Using Kepler's discoveries, Isaac Newton (1643–1727) deduced his *Law of Universal Gravitation (see page 88)*, and then went on to deduce his general laws of motion:

1. *An object at rest or in motion will remain in such a state until acted upon by some force.*

2. *The acceleration produced by a force on a given mass is proportional to that force.*

3. *A force exerted by A on B is always accompanied by an equal force exerted by B on A, in the same straight line but in the opposite direction.*

Albert Einstein (1879–1955) discovered that at velocities approaching that of light, Newton's Laws require significant modification.

Lower Speed

Equal areas,
Equal times

Higher
Speed

Kepler's Law of Areas

Force = mass × acceleration

An object falling to Earth
exerts an equal and opposite
gravitational pull on the
Earth itself. However, this is
rarely acknowledged as the
Earth, being of much greater
mass, experiences no
noticeable acceleration.

29

GRAVITY AND PROJECTILES
featherless falling objects

Newton's *Law of Universal Gravitation* states that any two masses m_1 and m_2 at a distance d apart will exert equal and opposite forces on each other of magnitude F where,

$$F = \frac{Gm_1 m_2}{d_2}$$

G is the *universal gravitation constant* (*see page 384*). Note that d is the distance between the *centres of mass* of m_1 and m_2.

Near the surface of the Earth (mass m_2), d is effectively constant, giving the 'local' gravitational *acceleration* constant g:

$$g = \frac{Gm_2}{d_2} = 9.80665 \text{ m/sec}^2 \text{ so that } F = m_1 g$$

Assuming negligible air resistance, an object falling from rest will travel s meters in t seconds where s and t are related by,

$$s = \tfrac{1}{2} g t^2 \quad \text{or} \quad t = \sqrt{\frac{2s}{g}}$$

The object's velocity at time t will be $v = gt = \sqrt{2gs}$ m/sec. Note that these quantities are independent of mass.

The path of a projectile with initial velocity v and angle of trajectory θ is given by,

$$x(t) = vt \cos\theta \quad \text{and} \quad y(t) = vt \sin\theta - \tfrac{1}{2} g t^2$$

These are time-dependent coordinates.

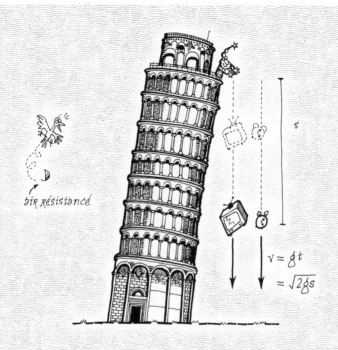

air resistance

s

$v = gt$
$= \sqrt{2gs}$

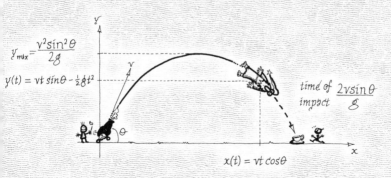

$y_{max} = \dfrac{v^2 \sin^2 \theta}{2g}$

$y(t) = vt \sin \theta - \frac{1}{2} g t^2$

time of impact $\dfrac{2v\sin\theta}{g}$

$x(t) = vt \cos \theta$

ENERGY, WORK, AND MOMENTUM
conservation in action

An object, mass m, moving in a straight line with velocity v has *kinetic energy* $E_k = \frac{1}{2}mv^2$. This is energy it possesses because of its motion. When an applied force changes its velocity to u, the total *work* done, W, is the change in kinetic energy: $W = \frac{1}{2}mv^2 - \frac{1}{2}mu^2$

Generally, *work* measures an exchange of energy between two bodies. Someone lifting an object of mass m to a height h above its initial position does work, here transferring gravitational *potential energy*, E_p, to the object (it can now fall). $E_p = mgh$ (mg is the *weight* of the object, a force).

When the object falls, it loses height but gains velocity, thus E_k increases as E_p decreases. Ignoring friction, the total energy of the object $E = E_k + E_p$ remains constant until it lands, when its remaining kinetic energy is dissipated as heat and noise.

The *linear momentum* of an object is given by $p = mv$. For a point mass m rotating about an axis at distance r, the *angular momentum*, L equals $(mv)r = (m\omega r)r = mr^2\omega$ where ω is the *angular velocity* of the body, in radians per second. $I = mr^2$ is known as the *moment of inertia*. The *rotational kinetic energy* of a system is then $E_{kr} = \frac{1}{2}I\omega^2$.

A general rotating solid can be treated as if it were a point mass rotating about the same axis with the appropriate *radius of gyration*. This is shown opposite and can be found using the methods of calculus (*see pages 50-51*). If no external forces act on a system, its total momentum is always conserved.

Kinetic energy is conserved so
$$\tfrac{1}{2}MV^2 = \tfrac{1}{2}Mp^2 + \tfrac{1}{2}mq^2$$
also because linear momentum is conserved:
$$Mp\sin\alpha - mq\sin\beta = 0$$
(horizontal component)
$$Mp\cos\alpha + mq\cos\beta = MV$$
(vertical component)

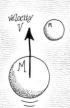

velocity V

M

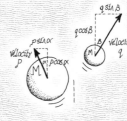

force = Rate of change of Momentum

The moment of inertia, I of a body rotating about an axis is given as
$$I = \sum mr^2 = \int r^2 \, dm$$

Torque results in angular acceleration α such that $T = I\alpha$

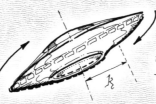

$$I = mK^2$$
where M is the total mass & K is the radius of gyration.

ROTATION AND BALANCE
whirling, gears, and pulleys

If an object of mass m on a string of length r is swung around in a circle at a velocity of v, then there is a *centripetal force*:

$$F = \frac{mv^2}{r} = m\omega^2 r \quad \text{giving acceleration} \quad a = \frac{v^2}{r} = \omega^2 r$$

toward the centre, where ω is the angular velocity. The centripetal force is equal and opposite to the string tension, which is also treated as a force.

Two interlocking gears with t_1 and t_2 teeth, and speeds r_1 and r_2 (in rpm or any other unit), relate by,

$$t_1\, r_1 = t_2\, r_2 \quad \text{or equivalently,} \quad r_1 = \frac{t_2}{t_1}\, r_2 \text{ and } r_2 = \frac{t_1}{t_2}\, r_1$$

The equation also holds true for two belt pulley wheels with diameters t_1 and t_2, and speeds r_1 and r_2.

If two objects with weights w_1 and w_2 are balanced, as shown opposite, at distances d_1 and d_2 from a fulcrum, then the *torque* associated with the two objects must be equal. Torque is the product of force and radial distance:

$$d_1\, w_1 = d_2\, w_2$$

A long-handled wrench turns a nut more easily than a short-handled one because it produces more torque.

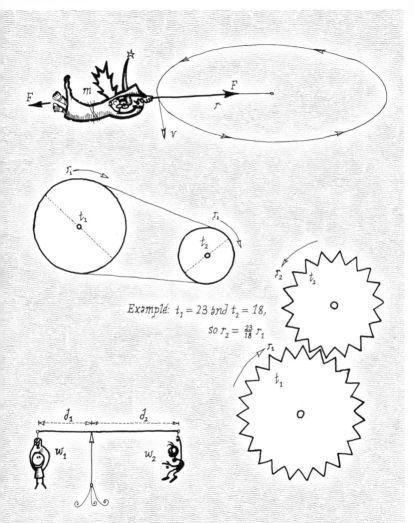

Example: $t_1 = 23$ and $t_2 = 18$,

so $r_2 = \frac{23}{18}\, r_1$

SIMPLE HARMONIC MOTION
vibrating and oscillating phenomena

Galileo Galilei (1564–1642) discovered that the *period T* of a pendulum, the time it takes to swing from one side to the other and back again, is independent of its *amplitude*, its maximum displacement from the centre. Thus a pendulum of length *l* completes *f* swings every second, regardless of whether its swings are wide or narrow. Here *f* is the *frequency* of the pendulum.

If *l* is given in meters, the period *T* is given in seconds by,

$$T = 2\pi \sqrt{\frac{l}{g}} = \frac{1}{f}$$

where *g* is the gravitational constant (*see page 56*).

For small swings, less than 5°, pendula approximate *simple harmonic motion*. Bobbing bottles and sproinging springs all perform *s.h.m.*, as do a vast array of vibrating and oscillating phenomena. Potential and kinetic energies are continually exchanged. The potential energy and acceleration of the vibrating object reach maxima when its kinetic energy and velocity are minimal and vice versa.

A simple way to generate *s.h.m.* is to project a uniform circular motion onto an axis as shown opposite, giving,

$$d(t) = a \sin \omega t$$

where *a* is the amplitude and *ω* the angular velocity. The period is then $\frac{2\pi}{\omega}$ seconds per cycle and the frequency is the reciprocal of the period, $\frac{\omega}{2\pi}$ cycles per second.

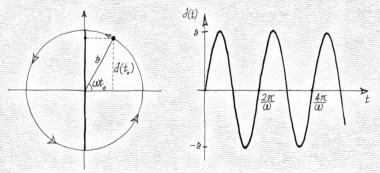

Projecting the uniform circular motion onto the vertical axis results in a non-uniform up-and-down motion directly related to the sine function.

STRESS, STRAIN, AND HEAT
expansion, contraction, tension, and compression

When a material is stretched or squashed it changes shape. The *stress*, σ, in the material is defined as the force F per unit area A; the *strain*, ε, is defined as the change in length, Δl, relative to the original length, l_0. *Young's Modulus E* for the material is then,

$$E = \frac{stress}{strain} = \frac{\sigma}{\varepsilon} = \frac{F/A}{\Delta l/l_0}$$

Substances also possess a *Bulk Modulus K*, relating inversely to its volumetric *compressibility*, and a *Shear Modulus G*, the ratio of shear stress and strain (*shown opposite*).

Heating and cooling cause expansion and contraction of materials in a linear relation to the temperature increase or decrease. The *linear expansivity*, α, of a substance relates its change in length Δl to its change in temperature ΔT, so that $\Delta l = \alpha l_0 \Delta T$, where l_0 is its starting length.

Hooke's Law states that a spring (or equivalently elastic object) stretched x units of distance beyond its equilibrium position pulls back with a force $F = kx$, where k is the *spring constant* of the spring in question. As a consequence, when a weight is attached to a vertical spring, the elongation is proportional to the weight.

For any structure to be in equilibrium, the total forces at any point must be balanced as shown opposite.

$$\text{Shear Stress} = \frac{\text{tangential force}}{\text{area}} = \frac{F}{A}$$

$$\text{Shear Strain} = \tan(\text{angle of shear } \gamma)$$

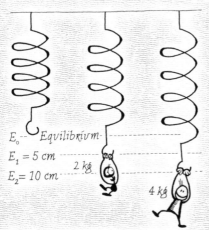

E_0 —— Equilibrium

$E_1 = 5$ cm

$E_2 = 10$ cm

2 kg

4 kg

If a 2 kg. force stretches a spring 5 cm, we can expect a 4 kg force to stretch it 10 cm. This law applies up to a certain point called the elastic limit.

At each point a balanced triangle of forces exists

LIQUIDS AND GASES
temperature, pressure, and flow

Liquid flowing at velocity v though a pipe with a cross-sectional area A has a rate of flow q where $q = Av$.

Liquid flowing through two pipes of cross-sectional areas A_1 and A_2, both subjected to the same pressure, will have flow velocities v_1 and v_2, where $A_1 v_1 = A_2 v_2$.

Pascal's Principle states that pressure applied to an enclosed liquid of any shape is transmitted evenly throughout, where pressure is defined as force per unit area. In the example illustrated opposite, $F_1 A_2 = F_2 A_1$.

Bernoulli's Equation states that a change of height results in a pressure change in a liquid: $p_1 + h_1 \rho g = p_2 + h_2 \rho g$ *(lower opposite)*.

Turning to gases, the *Perfect Gas Law* states that if a fixed quantity of gas has pressure P, volume V, and temperature T (in degrees *Kelvin*), then PV is proportional to T. The Kelvin scale is a measure of *absolute temperature*, where $°K = °C + 273.15$.

Given a closed system with initial pressure, volume, and temperature P_1, V_1, and T_1, and later values P_2, V_2, and T_2 :

Charles' Law states: $\dfrac{V_1}{V_2} = \dfrac{T_1}{T_2}$ for a constant pressure,

Boyle's Law states: $P_1 V_1 = P_2 V_2$ for a constant temperature.

$$F_1 \qquad F_2 = \frac{A_2}{A_1} F_1$$

Pressure applied to an enclosed liquid of any shape is transmitted evenly throughout. This is Pascal's principle, the basis of hydraulics.

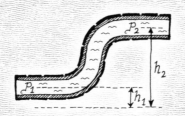

Here ρ denotes the average density of the fluid, given by mass/volume. If we are using the metric unit g/cm^3, then water gives $\rho = 1$. g is the gravitational constant

41

SOUND

harmonic wavelengths and passing sirens

For a string fixed at two points distance L apart, as shown, the harmonic wavelengths which fit the string are given by,

$$\lambda_n = \frac{2L}{n} \quad \text{for } n = 1, 2, 3 \ldots$$

Only λ_1 and the first few *overtones* may actually be audible.

If W_T is the wave velocity in the string, T its tension (in newtons), and μ the mass per unit length (in kg/m), then the basic frequencies, v_n, (in cycles per second) are,

$$v_n = \frac{W_T}{\lambda_n} = \frac{n}{2L} W_T = \frac{n}{2L}\sqrt{\frac{T}{\mu}}$$

The tone of an ambulance siren appears to change as it passes. This is known as the *Döppler Effect*. If an observer is moving at velocity v_o *toward* a source of frequency f_s, which is itself moving at velocity v_s *toward* the observer, then for a wave of speed c the observed frequency f_o is,

$$f_o = \left(\frac{c + v_o}{c - v_s}\right) f_s$$

Note: if objects recede rather than approach, use negative values for v_o and v_s (the speed of sound is 331.45 m/sec).

If two very similar tones, frequencies f_1 and f_2, are played together, a *beat frequency* $f_{beat} = (f_2 - f_1)$ may be heard.

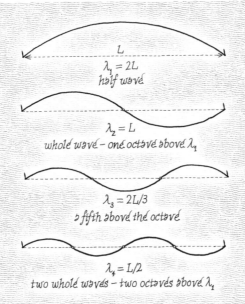

$\lambda_1 = 2L$
half wave

$\lambda_2 = L$
whole wave - one octave above λ_1

$\lambda_3 = 2L/3$
a fifth above the octave

$\lambda_4 = L/2$
two whole waves - two octaves above λ_1

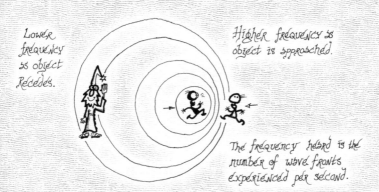

Lower frequency as object Recedes.

Higher frequency as object is approached.

The frequency heard is the number of wave fronts experienced per second.

43

LIGHT

refraction, lenses, and relativity

Light travelling from air into water is shown opposite. If v_1 and v_2 are the speeds that light takes in any two media then *Snell's Law of Refraction* states that for a given frequency:

$$\frac{v_1}{v_2} = \frac{\sin\theta_1}{\sin\theta_2} = \text{constant} \quad \text{or} \quad n_1\sin\theta_1 = n_2\sin\theta_2$$

where n_1 and n_2 are the *refractive indices* of the two media (varying slightly for different frequencies). Here are a few useful values:

Vacuum and air: 1   *Water*: 1.33   *Quartz*: 1.45   *Crown glass*: 1.52

Various lenses are shown opposite, some converging, others diverging. The *focal length, f*, is shown for one. The *Gaussian Lens Equation* relates the distances between an object, the lens, and the upside-down focused image:

$$\frac{1}{x_o} + \frac{1}{x_i} = \left(\frac{n_{lens}}{n_{medium}} - 1\right)\left(\frac{1}{R_1} + \frac{1}{R_2}\right) = \frac{1}{f}$$

Here R_1 and R_2 are the left and right *radii of curvature* of the lens (negative if concave).

Visible light is a tiny part of the *electromagnetic spectrum* which also includes x-rays, radio, and microwaves. Einstein was able to deduce that, because light travels at a constant speed away from you irrespective of your own speed, time itself must be able to stretch and dilate! This is part of his *Special Theory of Relativity*.

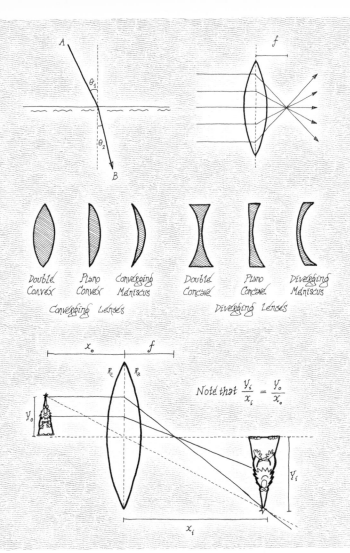

A

θ_1

θ_2

B

f

Double
Convex

Plano
Convex

Converging
Meniscus

Converging Lenses

Double
Concave

Plano
Concave

Diverging
Meniscus

Diverging Lenses

x_o

f

R_1 R_2

Note that $\dfrac{y_i}{x_i} = \dfrac{y_o}{x_o}$

y_o

y_i

x_i

ELECTRICITY AND CHARGE
tuning the circuit

For a simple electrical circuit, the *voltage E* (in volts) across a *resistance R* (in ohms, Ω) produces a *current I* (in amps) defined by *Ohm's Law*, $E = IR$. The *power P* (in watts) in the circuit is then,

$$P = EI = I^2 R$$

Resistors in series give resistance $R_s = R_1 + \ldots + R_n$ ohms. Capacitors in parallel give capacitance $C_p = C_1 + \ldots + C_n$ faradays. Resistors in parallel and capacitors in series combine to give,

$$R_p = \frac{1}{\frac{1}{R_1} + \ldots + \frac{1}{R_n}} \qquad C_s = \frac{1}{\frac{1}{C_1} + \ldots + \frac{1}{C_n}}$$

Equations for circuits involving inductors are shown opposite.

Electrical effects result from *charge* (in coulombs). The charge of an electron is -1.6×10^{-19} C. *Coulomb's Law* states that the force *F* between two point charges Q_1 and Q_2 separated by distance *r* meters is given by,

$$F = \frac{Q_1 Q_2}{4\pi\varepsilon_0 r^2}$$

where ε_0 is the *permittivity of empty space*, 8.85×10^{-12} farad/m.

A *diode* acts as a one-way valve for current, while *transistors* are more like a tap or faucet, where a voltage at the base (*B*) controls the current flow from the collector (*C*) to the emitter (*E*). They can then be used as switches or amplifiers (*see simple amplifier opposite*).

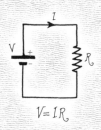

$$V = IR$$

Energy Stored

in an Inductor
$$= \frac{1}{2}LI^2$$

in a Capacitor
$$= \frac{1}{2}CV^2$$

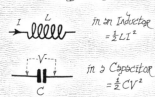

Parallel Resonant Circuit

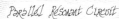

Variable Capacitor

Voltage peaks with respect to tuned frequency.

tuned frequency
$$= \frac{1}{2\pi\sqrt{LC}}$$

Series Resonant Circuit

Current peaks with respect to tuned frequency.

Voltage gain is given by

$$A_v = -\frac{R_C}{R_E} = \frac{V_{out}}{V_{in}}$$

R1 and R2 are bias resistors where

$$V_B = \frac{V_{cc} R_2}{R_1 + R_2}$$

C1 and C2 are coupling capacitors

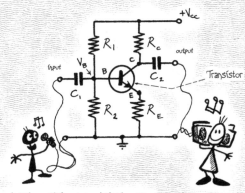

Simple common emitter amplifier using an NPN Transistor

ELECTROMAGNETIC FIELDS
charge, flux, and handedness

Charge behaves in an electromagnetic field similarly to mass in a gravitational field. The force **F** on a moving charge Q in an electromagnetic field of strength **E** is given by **F** = **E**Q. The force **F** on a wire carrying a current I and with length l is **F** = **B**Il, where **B** is the *magnetic flux density*, measured in *teslas*. Other situations are shown opposite (μ_0 is the permeability of free space, the magnetic constant). Note: boldfaced quantities are vectors, with directions as well as magnitude.

A point charge Q moving with velocity v creates a magnetic field **B** (*opposite top left*). If point P has distance r from Q and is at angle θ to its direction of motion, then the *Biot-Savart Law* states that

$$\mathbf{B} = \left(\frac{\mu_0}{4\pi}\right)\frac{Qv\sin\theta}{r^2}$$

A moving magnetic field produces an electric field and vice versa. Using magnets and currents in coils of wire electrical energy can be transformed into mechanical energy (a *motor*) and vice versa (a *dynamo*). Fleming's left- and right-hand rules apply respectively (*lower opposite*).

Often useful in designing such things are *Faraday's Law of Induction*, (one of Maxwell's four equations) which states that the induced electromotive force in any closed circuit is proportional to the rate of change of the magnetic flux, or $\nabla \times \mathbf{E} = -\frac{\partial \mathbf{B}}{\partial t}$, where $\nabla \times$ is the *curl operator*, and *Lenz's Law*, which states that an induced current is always in such a direction as to oppose the motion or change causing it.

current is in direction of movement of positive charge.

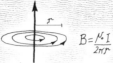

$$B = \frac{\mu_0 I}{2\pi r}$$

field strength at distance r from a wire

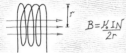

$$B = \frac{\mu_0 I N}{2r}$$

for a coil radius r with N turns

$$B = \mu_0 \frac{N}{l} I$$

for a spiral with N turns in length l.

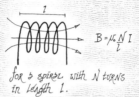

PULSED D.C. DYNAMO (water powered)

Torque = ABIN
where A is area of coil

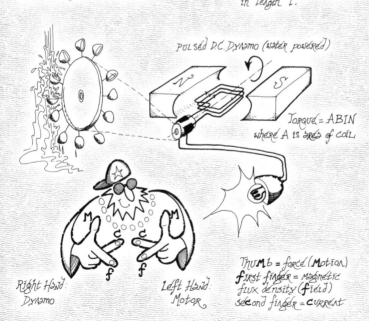

Right Hand: Dynamo

Left Hand: Motor

THUMB = force (MOTION)
First finger = magnetic flux density (Field)
seCond finger = Current

CALCULUS
differentiation and integration

Calculus makes use of *infinitesimals* and *limits* to solve two problems, the instantaneous rate of change of a function and the exact area under a curve.

The graph of the function $y = f(x)$ at position $(a, f(a))$ has gradient $f'(a)$, where the function $f'(x) = \frac{df}{dx}(x)$ is called the *derivative* of f. For each $x, f'(x)$ is the rate of change of f at x. To directly calculate such a derivative at a, we consider the slopes of the lines through $(a, f(a))$ and $(a+\varepsilon, f(a+\varepsilon))$ for ever-tinier values of ε. If they tend toward a 'limit', then the rate of change of f at a can be defined to be this limit.

If $x(t)$ denotes the position of an object at time t, then its velocity $v(t)$ at time t is $x'(t) = \frac{dx}{dt}(t)$. Its acceleration $a(t)$, the rate of change of its velocity at t, is then $x''(t) = \frac{dv}{dt}(t) = \frac{d^2x}{dt^2}(t)$.

Suppose we have a function (*opposite*) and seek the area beneath its graph between a and b. The interval between a and b is divided into an ever-larger number of equal lengths. This produces an ever-larger number of narrowing rectangles, the sum of whose areas can easily be found at each stage. The area under the curve is given by the limit of these sums, written $\int_a^b f(x)dx$. If $F(x)$ satisfies $F'(x) = f(x)$, then remarkably, $\int_a^b f(x)dx = F(b) - F(a)$.

$F(x)$ is called the *indefinite integral* or *antiderivative* of f and denoted $\int f\,dx$. As $(F(x) + c)' = F'(x)$, it follows that all of the antiderivatives given opposite involve an arbitrary constant.

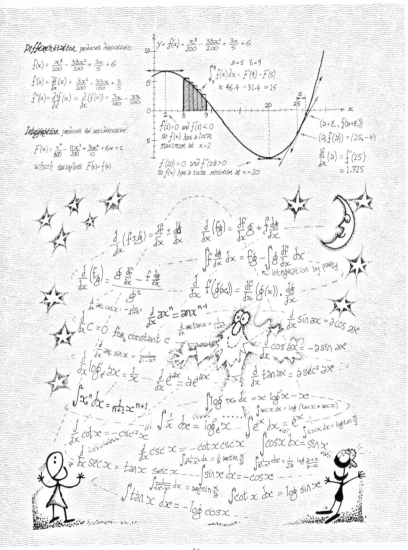

Differentiation produces derivatives:

$$f(x) = \frac{x^3}{200} - \frac{33x^2}{200} + \frac{3x}{5} + 6$$

$$f'(x) = \frac{df}{dx}(x) = \frac{3x^2}{200} - \frac{33x}{100} + \frac{3}{5}$$

$$f''(x) = \frac{d^2f}{dx^2}(x) = \frac{d}{dx}\left(f'(x)\right) = \frac{3x}{100} - \frac{33}{100}$$

Integration produces the antiderivative:

$$F(x) = \frac{x^4}{800} - \frac{11x^3}{200} + \frac{3x^2}{10} + 6x + c$$

which satisfies $F'(x) = f(x)$

$$y = f(x) = \frac{x^3}{200} - \frac{33x^2}{200} + \frac{3x}{5} + 6$$

$a=5\quad b=9$

$$\int_5^9 f(x)\,dx = F(9) - F(5) \approx 46.4 - 31.4 = 15$$

$f'(2) = 0$ and $f''(2) < 0$ so $f(x)$ has a local maximum at $x=2$

$f'(20) = 0$ and $f''(20) > 0$ so $f(x)$ has a local minimum at $x=20$

$(2+\varepsilon, f(2+\varepsilon))$

$(2, f(2)) = (25, -4)$

$$\frac{df}{dx}(2) = f'(25) = 1.725$$

$$\frac{d}{dx}(f+g) = \frac{df}{dx} + \frac{dg}{dx} \qquad \frac{d}{dx}(fg) = \frac{df}{dx}g + f\frac{dg}{dx}$$

$$\int f\frac{dg}{dx}\,dx = fg - \int g\frac{df}{dx}\,dx \quad \text{integration by parts}$$

$$\frac{d}{dx}\left(\frac{f}{g}\right) = \frac{g\frac{df}{dx} - f\frac{dg}{dx}}{g^2} \qquad \frac{d}{dx}f(g(x)) = \frac{dF}{dx}(g(x)) \cdot \frac{dg}{dx}$$

$$\frac{d}{dx}\arccos x = -\frac{1}{\sqrt{1-x^2}}$$

$$\frac{d}{dx}ax^n = anx^{n-1} \qquad \frac{d}{dx}\arctan x = \frac{1}{1+x^2} \qquad \frac{d}{dx}\sin ax = a\cos ax$$

$$\frac{d}{dx}C = 0 \text{ for constant } c$$

$$\frac{d}{dx}\arcsin x = \frac{1}{\sqrt{1-x^2}} \qquad \frac{d}{dx}\cos ax = -a\sin ax$$

$$\frac{d}{dx}\log_e x = \frac{1}{x} \qquad \frac{d}{dx}e^{ax} = ae^{ax} \qquad \frac{d}{dx}\tan ax = a\sec^2 ax$$

$$\int x^n\,dx = \frac{1}{n+1}x^{n+1} \qquad \int \log_e x\,dx = x\log_e x - x$$

$$\int \frac{1}{x}\,dx = \log_e x \quad \cdots\cdots \int e^x\,dx = e^x \qquad \int \sec x\,dx = \log(\tan x + \sec x)$$

$$\int \csc x\,dx = \log\tan\frac{x}{2}$$

$$\frac{d}{dx}\cot x = -\csc^2 x \qquad \int \cos x\,dx = \sin x$$

$$\frac{d}{dx}\csc x = -\cot x\,\csc x$$

$$\frac{d}{dx}\sec x = \tan x\,\sec x \qquad \int \frac{1}{a^2+x^2}\,dx = \frac{1}{a}\arctan\frac{x}{a}$$

$$\int \frac{1}{a^2-x^2}\,dx = \frac{1}{2a}\log\frac{a+x}{a-x} \qquad \int \sin x\,dx = -\cos x$$

$$\int \frac{1}{\sqrt{a^2-x^2}}\,dx = \arcsin\frac{x}{a} \qquad \int \cot x\,dx = \log\sin x$$

$$\int \tan x\,dx = -\log\cos x$$

COMPLEX NUMBERS

into the imaginary realm

The familiar *real numbers* are contained within the larger realm of *complex* numbers. These are constructed by starting with the *imaginary unit* which is denoted i, and which satisfies (unlike any real number):

$$i^2 = -1 \text{ or } i = \sqrt{-1}$$

Given any two real numbers a and b, the quantity $a + bi$ is called a complex number. Some complex equations are shown below:

$$a + bi = c + di \text{ if and only if } a = c \text{ and } b = d$$

$$(a + bi) + (c + di) = (a + c) + (b + d)i$$

$$(a + bi)(c + di) = (ac - bd) + (ad + bc)i$$

$$\frac{ac + bd}{c^2 + d^2} = \frac{a + bi}{c + di} + \frac{bc - ad}{c^2 + d^2} i$$

Polar representation uses the angle θ and radius r:

$$z = r\cos\theta + ir\sin\theta = r(\cos\theta + i\sin\theta)$$

The exponential function e^x may be extended into the complex plane using *Euler's equation*,

$$e^{i\theta} = \cos\theta + i\sin\theta$$

From this we have both the mathematical gem $e^{i\pi} = -1$, and *DeMoivre's Theorem*, for powers of a complex number z:

$$z^n = (re^{i\theta})^n = r^n e^{in\theta} = r^n(\cos n\theta + i\sin n\theta)$$

The real number line contains not only the positive and negative whole numbers, but all positive and negative fractions and irrational numbers such as $\sqrt{2}$ and π.

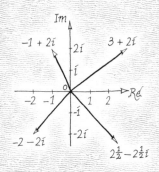

The real number line becomes the Re(al) axis in the plane of complex numbers. Each point in the plane corresponds to a complex number, and vice versa. Numbers on the Im(aginary) axis are called 'pure imaginary', having a real component of 0.

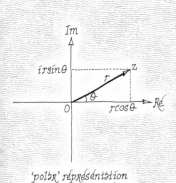

'polar' representation of a complex number.

The complex plane is the home of the famed fractal Mandelbrot Set, which can be generated using a process involving the iteration of the map $z \rightarrow z^2 + c$.

RELATIVITY THEORY
length contractions and time dilations

Einstein extended Galileo's principle of relativity of uniform motion from mechanics to all laws of physics (including electrodynamics), while incorporating the constant nature of the speed of light, c.

At the core of his *Special Theory of Relativity* are the *Lorentz transformations* relating a coordinate frame (x', y', z', t') to another, (x, y, z, t), which it is moving relative to at velocity v. In the special case where motion is parallel to (for example) the x-axis, these are:

$$x' = \gamma(x - vt), \ \ y' = y, \ \ z' = z, \ \ t' = \gamma(t - vx/c^2) \ \text{ where } \ \gamma = \left(1 - v^2/c^2\right)^{-1/2}$$

Note that these transformations result in length contraction in the x-direction *and* time dilation (but both are negligible when v is small relative to c). For a more general direction of motion, a matrix-based equation provides the analogous transformations.

Composing velocities u and v in this context gives not $u + v$, but,

$$\frac{u + v}{1 + \left(\frac{uv}{c^2}\right)}$$

The (scalar) energy and the momentum (vector) of an object with mass m and velocity (vector) v are given, respectively, by

$$E = \gamma mc^2, \ \ p = \gamma mv$$

and if the object is at rest, we get that most famous final equation,

$$E = mc^2$$

EINSTEIN'S field equation

$$R_{\mu\nu} - \frac{1}{2}g_{\mu\nu}R + g_{\mu\nu}\Lambda = \frac{8\pi G}{c^4} T_{\mu\nu}$$

Newton's gravitational constant

curvature tensor   metric tensor   cosmological constant

speed of light

stress energy tensor (describes distribution of matter and energy)

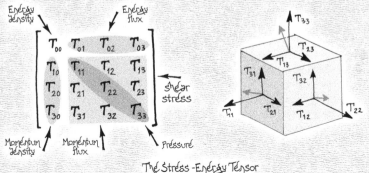

The Stress-Energy Tensor

CONSTANTS

Speed of Light	c	670,616,629 mph, 186,282.4 miles s^{-1}, 2.99792458 x 10^8 m s^{-1}
Light-year	ly	0.306601 pc, 5,878,625,373,183.608 miles, 9,460,730,472,580 km
Parsec	pc	3.26156 ly, 1.9173508 x 10^{13} miles, 3.085678 x 10^{13} km
Astronomical Unit	AU	1.5813 x 10^{-5} ly, 92,955,791 miles, 149,597,892 km
Speed of Sound		*in dry air at 0°C* 331.4 m s^{-1}, 1087.3 feet s^{-1} *in water at 20°C* 1482 m s^{-1}, 4862 feet s^{-1}

Electron Mass	m_e	9.1093897 × 10^{-31} kg	*Gas constant*	R	8.314472 J K^{-1} mol^{-1}
		= 0.51099 MeV	*Gravitational const.*	G	6.674259 x 10^{-11} m^3 kg^{-1} s^{-2}
Electron charge	e	1.602176 × 10^{-19} C	*H₂0 triple point*	T_{tpw}	273.16 K
		= 4.803 × 10^{-10} esu	*Imped. of vacuum*	Z_o	376.7303 Ω
Proton mass	m_p	1.672623 × 10^{-27} kg	*Loschmidt const.*	n_o	2.6867775 x 10^{25} m^{-3}
		= 1836.1 x electron mass	*Mag. flux quantum*	ϕ_o	2.067833758 x 10^{-15} Wb
Neutron mass	m_n	1.6749286 x 10^{-27} kg	*Magnetic const.*	μ_o	4π x 10^{-7} NA^{-2}
Atomic mass unit	u	1.66054 x 10^{-27} kg	*Mol. vol. gas*	V_m	2.2413968 x 10^{-2} m^3 mol^{-1}
Atm. mass eng. eqv.	$m_u c^2$	931.494 MeV	*Perm. of vacuum*	μ_o	4π x 10^{-7} H m^{-1}
Avogadro's no.	N_A	6.02214 x 10^{23} mol^{-1}	*Permittivity const.*	ε_o	8.8541878 x 10^{-12} F m^{-1}
Bohr radius	a_o	5.2917725 x 10^{-11} mol^{-1}	*Planck constant*	h	6.62606896 x 10^{-34} J s
Boltzmann const.	k	1.38065 x 10^{-23} J K^{-1}	*Planck const/2π*	h	1.054571726 x 10^{-34} J s
Earth Escape Veloc.	v_{esc}	6.96 miles s^{-1}, 11.2 kms^{-1}	*Planck length*	l_P	1.616199 x 10^{-35} m
Earth Grav. Accel.	g_n	32.174 feet s^{-2},	*Planck time*	t_P	5.31906 x 10^{-44} s
		9.80665 ms^{-2}	*Planck temp*	T_P	1.41683 x 10^{32} K
Earth Std. Atmos.		101,325 Pa	*Planck mass*	m_P	2.17651 x 10^{-8} kg
Faraday constant	F	9.648534 x 10^4 C mol^{-1}	*Rcp. fine str. const.*	$1/\alpha$	137.035999
Fine struct. const.	α	7.297353 x 10^{-3}	*Rydberg const.*	R_H	1.097373 x 10^7 m^{-1}

UNITS

Physical Quantity	Symbol for quantity	Name of SI unit	Unit symbol	Immediate definition	Basic units definition	As mass, length, time, & current
frequency	f	hertz	Hz	s^{-1}	s^{-1}	T^{-1}
force	F	newton	N	kg m s^{-2}	kg m s^{-2}	M L T^{-2}
energy	W	joule	J	N m	kg m^2 s^{-2}	M L^2 T^{-2}
power	P	watt	W	J s^{-1}	kg m^2 s^{-3}	M L^2 T^{-3}
pressure	p	pascal	Pa	N m^{-2}	kg m^{-1} s^{-2}	M L^{-1} T^{-2}
charge	Q	coulomb	C	A s	A s	T I
voltage	V	volt	V	J C^{-1}	kg m^2 s^{-3} A^{-1}	M L^2 T^{-3} I^{-1}
capacitance	C	faraday	F	C V^{-1}	A^2 s^4 kg^{-1} m^{-2}	M^{-1} L^{-2} T^4 I^2
resistance	R	ohm	Ω	V A^{-1}	kg m^2 s^{-3} A^{-2}	M L^2 T^{-3} I^{-2}
conductance	G	siemens	S	Ω^{-1}	kg^{-1} m^{-2} s^3 A^2	M^{-1} L^{-2} T^3 I^2
flux density	B	tesla	T	N A^{-1} m^{-1}	kg s^{-2} A^{-1}	M T^{-2} I^{-1}
magnetic flux	Φ	weber	Wb	T m^2	kg m^2 s^{-2} A^{-1}	M L^2 T^{-2} I^{-1}
inductance	L	henry	H	V A^{-1} s	kg m^2 s^{-2} A^{-2}	M L^2 T^{-2} I^{-2}

WEIGHTS & MEASURES

British Standard Imperial Weights & Measures

1 yard = 3 feet = 36 inches

1 furlong = 220 yards = 660 feet

1 nautical mile = 6080 feet

1 pound = 16 ounces

1 hundredweight = 112 pounds

1 pint = 4 gills = 20 fluid ounces

1 fathom = 2 yards = 6 feet

1 mile = 1760 yards = 5280 feet

1 acre = 4840 square yards

1 stone = 14 pounds

1 ton = 2240 pounds

1 gallon = 277.4 cubic inches

British Imperial : Metric

1 inch = 2.54000 cm

1 foot = 0.304800 meter

1 mile = 1.60934 kilometers

1 ounce = 28.3495 grammes

1 pound = 0.45359237 kilogram

1 (long) ton = 1016.047 kg

1 gallon = 4.549631 litres

1 bushel = 36.397 litres

1 acre = 0.404687 hectares

1 cubic inch = 16.3871 cubic cm

$^{\circ}$fahrenheit = $\frac{9}{5}$($^{\circ}$celsius) + 32

1 horsepower = 0.7457 kilowatts

1 pound per sq. inch = 0.0688 atm

1 foot pound = 1.355 joules

Metric : British Imperial

1 cm = 0.393701 inches

1 meter = 3.280842 feet

1 kilometer = 0.621371 miles

1 gramme = 0.0352740 ounces

1 kilogram = 2.204622 pounds

1 tonne = 0.9842064 (long) tons

1 litre = 0.219975 gallons

1 litre = 0.027475 bushels

1 hectare = 2.47105 acres

1 cubic cm = 0.0610237 cubic inches

$^{\circ}$celsius = $\frac{5}{9}$($^{\circ}$fahrenheit − 32)

1 kilowatt = 1.3410 horsepower

1 atmosphere = 14.696 lbs/sq. inch

1 joule = 0.738 foot pounds

US Weights & Measures differing from UK

1 US hundredweight = 100 pounds

1 US (short) ton = 0.892857 UK (long) tons

1 US gallon = 3.785412 litres

1 US gallon = 0.83267 UK gallons

1 dry gallon = 4.404884 litres

1 US bushel = 35.2391 litres

1 US bushel = 0.9689 UK bushels

1 US (short) ton = 2000 pounds

1 US (short) ton = 907.184 kilogram

1 litre = 0.264172 US gallons

1 UK gallon = 1.20095 US gallons

1 litre = 0.227021 dry gallons

1 litre = 0.028378 US bushels

1 UK bushel = 1.0321 U.S. bushels

General Metric

$^{\circ}$kelvin (K) = $^{\circ}$celsius + 273.15

1 atmosphere = 101325 Pa

1 calorie (cal) = 4.184 J

1 esu = 3.3356 × 10^{-10} C (coulombs)

1 erg = 2.390 × 10^{-11} kcal

1 eV / molecule = 96.485 kJ mol^{-1}

1 kcal mol^{-1} = 349.76 cm^{-1}, 0.0433 eV

1 newton per square meter (N m^{-2}) = 1 Pascal (Pa)

1 angström (A) = 10 × 10^{-10} m

1 bar = 105 Pa

1 curie (Ci) = 3.7 × 10^{10} s^{-1}

1 erg = 10^{-7} J

1 eV = 1.60218 × 10^{-19} J

1 joule (J) = 0.2389 cal

1 kJ mol^{-1} = 83.54 cm^{-1}

1 wave no. (cm^{-1}) = 2.8591 × 10^{-3} kcal mol^{-1}

Prefixes: 10^{3} = kilo 10^{6} = mega 10^{9} = giga 10^{12} = tera 10^{15} = peta 10^{18} = exa

10^{-3} = milli 10^{-6} = micro 10^{-9} = nano 10^{-12} = pico 10^{-15} = femto 10^{-18} = atto

EXPANSIONS & EXTRAS

$\pi = 3.14159265358979\ldots$ $\qquad$ $e = 2.718281828459045\ldots$

$\sqrt{2} = 1.414213562373095\ldots$ $\qquad$ $\sqrt{\pi} = 1.7724538500905516\ldots$

$e^x = 1 + x + \dfrac{x^2}{2!} + \dfrac{x^3}{3!} + \dfrac{x^4}{4!} + \ldots, \quad$ so $\quad e = 1 + 1 + \dfrac{1}{2!} + \dfrac{1}{3!} + \dfrac{1}{4!} + \ldots$

$\log(1+x) = x - \dfrac{x^2}{2} + \dfrac{x^3}{3} - \dfrac{x^4}{4} + \ldots \ (-1 < x < 1)$ $\qquad$ $e = 2 + \dfrac{1}{1 + \dfrac{1}{2 + \dfrac{1}{1 + \dfrac{1}{1 + \dfrac{1}{4 + \dfrac{1}{1 + \ddots}}}}}}$

$\sqrt{2} = 1 + \dfrac{1}{2 + \dfrac{1}{2 + \dfrac{1}{2 + \dfrac{1}{2 + \dfrac{1}{2 + \ddots}}}}}$ $\qquad$ $\sqrt{3} = 1 + \dfrac{1}{1 + \dfrac{1}{2 + \dfrac{1}{1 + \dfrac{1}{2 + \ddots}}}}$ $\qquad$ $\emptyset = 1 + \dfrac{1}{1 + \dfrac{1}{1 + \dfrac{1}{1 + \dfrac{1}{1 + \ddots}}}}$

$\dfrac{1}{1-x} = 1 + x + x^2 + x^3 + x^4 + \ldots \ (-1 < x < 1)$

$\pi = 4\left(\dfrac{1}{1} - \dfrac{1}{3} + \dfrac{1}{5} - \dfrac{1}{7} + \dfrac{1}{9} - \ldots\right)$ $\qquad$ $\pi = 3 + \dfrac{1^2}{6 + \dfrac{3^2}{6 + \dfrac{5^2}{6 + \dfrac{7^2}{6 + \dfrac{9^2}{6 + \dfrac{11^2}{6 + \ddots}}}}}}$

$\arcsin x = x + \dfrac{1}{2}\dfrac{x^3}{3} + \dfrac{1}{2}\dfrac{3}{4}\dfrac{x^5}{5} + \dfrac{1}{2}\dfrac{3}{4}\dfrac{5}{6}\dfrac{x^7}{7} + \ldots \quad$ *radians*

$\sin x = x - \dfrac{x^3}{3!} + \dfrac{x^5}{5!} - \dfrac{x^7}{7!} + \ldots, \quad \cos x = 1 - \dfrac{x^2}{2!} + \dfrac{x^4}{4!} - \dfrac{x^6}{6!} + \ldots \quad$ (*x in radians*)

Taylor expansion: $f(x) = f(x-a) + af'(x-a) + \dfrac{a^2}{2!}f''(x-a) + \dfrac{a^3}{3!}f'''(x-a) + \ldots$

Maclaurin expansion: $f(x) = f(0) + xf'(0) + \dfrac{x^2}{2!}f''(0) + \dfrac{x^3}{3!}f'''(0) + \ldots$

Fractions: $\quad \dfrac{a}{b} + \dfrac{c}{d} = \dfrac{ad+bc}{bd} \quad \dfrac{a}{b} - \dfrac{c}{d} = \dfrac{ad-bc}{bd} \quad \dfrac{a}{b} \times \dfrac{c}{d} = \dfrac{ac}{bd} \quad \dfrac{a}{b} \div \dfrac{c}{d} = \dfrac{ad}{bc}$

'*Tan $\frac{1}{2}$*' *formulæ*: $\quad$ if $t = \tan\frac{1}{2}\theta, \quad \sin\theta = \dfrac{2t}{1+t^2} \quad \cos\theta = \dfrac{1-t^2}{1+t^2} \quad \tan\theta = \dfrac{2t}{1-t^2}$

Riemann's zeta function: "*Probably the most challenging and mysterious object of modern mathematics, in spite of its utter simplicity*" M.C. Gutzwiller

$\zeta(x) = 1 + \dfrac{1}{2^x} + \dfrac{1}{3^x} + \dfrac{1}{4^x} \ldots = \left(\dfrac{1}{1-\frac{1}{2^x}}\right)\left(\dfrac{1}{1-\frac{1}{3^x}}\right)\left(\dfrac{1}{1-\frac{1}{5^x}}\right).\quad\left(\dfrac{1}{1-\frac{1}{p_k^x}}\right) \quad (x > 1)$

where p_k is the kth prime number.